Reading Essentials® in Science

COMMUNITIES OF LIFE

Tropical Rain Forests

JANE HURWITZ

PERFECTION LEARNING®

Editorial Director:	Susan C. Thies
Editor:	Mary L. Bush
Design Director:	Randy Messer
Book Design:	Brianne Osborn, Emily J. Greazel
Cover Design:	Michael A. Aspengren

A special thanks to the following for their scientific review of the book:
Paul Pistek, Instructor of Biological Sciences, North Iowa Area Community College
Jeffrey Bush, Field Engineer, Vessco, Inc.

For Devora, may she never find herself caught out on a limb

Image Credits:
© Buddy Mays/CORBIS: pp. 4–5, © John Holmes; Frank Lane Picture Agency/CORBIS: p. 10; © Chris Hellier/CORBIS: p. 14 (bottom); © Pablo Corral Vega/CORBIS: pp. 16–17; © Philip Gould/CORBIS: p. 19; © Michael & Patricia Fogden/CORBIS: p 21; © Douglas Faulkner/CORBIS: pp. 22–23; © Tim Davis/CORBIS: p. 25 (top); © Wendy Stone/CORBIS: p. 27; © D. Boone/CORBIS: pp. 28–29; © Wayne Lawler; Ecoscene/CORBIS: p. 30

Corel Professional Photos: cover, back cover, pp. 1, 3, 6 (bottom), 7, 8–9, 12–13, 18, 20, 25 (bottom), 26, 32, 33, 35, 38, 39, 40; MapArt: pp. 11, 13, 14 (middle), 15; Perfection Learning Corporation: pp. 17, 31; Photos.com: pp. 6 (top), 24, 34, 37

For information, contact
Perfection Learning® Corporation
1000 North Second Avenue, P.O. Box 500
Logan, Iowa 51546-0500.
Phone: 1-800-831-4190
Fax: 1-800-543-2745
perfectionlearning.com

2 3 4 5 6 7 PP 09 08 07 06 05 04

Paperback ISBN 0-7891-6058-7
Reinforced Library Binding ISBN 0-7569-4184-9

Contents

Introduction

If someone asked you to describe the area where you live, what would you say? Do you live in a desert region where it's hot and dry? a forest area with lots of evergreen trees? near a hot, wet tropical rain forest? How would you describe the temperature, sunlight, and rainfall in your hometown? What plants and animals live there?

Biomes

What you are describing is a biome. A **biome** is an **environment** with unique features. For example, an ocean biome has salt water. A **tundra** biome is cold and dry, and often the ground is frozen year-round.

There are many types of biomes, including desert, mountain, tundra, forest, grassland, ocean (saltwater), freshwater, and rain forest. Ecologists have noticed that the same biomes can appear in very different places. Deserts, for example, are found in both hot and cold locations. But even though they are in different parts of the world, all deserts share some characteristics.

Biome Career

Ecologists are scientists who study the relationship between the Earth and the living things on the planet.

Each biome has its own special plant life. Think about the different plants found in a desert, a rain forest, and a grassland. Cactuses grow in the desert. Palm trees grow in the rain forest. A variety of grasses cover the grassland.

Biomes are also identified by how plants and animals must **adapt** in order to live there. To live in an ocean biome, plants and animals must be able to live in salt water. In a desert, the wildlife must be able to survive long periods without water. Each biome has its own unique environment to which the plants and animals must adapt.

Coyotes adapt to life in the desert by growing lighter, thinner, shorter coats. They also have the ability to pant, which helps cool their body temperatures in desert heat.

Ecosystems

Ecologists have also determined that certain groups of plants and animals tend to live together. These groups of living creatures interact with the nonliving parts of the environment, such as rocks or sand. Groups of living creatures that interact with one another and their surroundings are called **ecosystems**.

Each biome is made up of many ecosystems. In an ocean, there are different ecosystems living in **coral reefs**, cold Arctic waters, and deep underwater **trenches**. Each layer of a tropical rain forest has its own ecosystems.

Working together, the ecosystems form **communities** of life within each of the biomes.

Fish gather near coral reefs to feed on the plants and animals living there.

Chapter 1

The Tropical Rain Forest

The tropical rain forest biome is a community of wildlife living in a moist, warm **climate**. Only certain areas in the world have the ideal temperature, sunlight, and rainfall needed for a tropical rain forest to form. Many of the plants and animals that live in tropical rain forests are found nowhere else in the world.

What Is a Tropical Rain Forest?

Forests can be found all over the world. A forest is an area of land covered with trees and **underbrush**. There are different types of forests, each with its own characteristics. A tropical rain forest has several special qualities.

Tropical rain forests grow near the **equator**. This area receives a steady amount of sunlight throughout the year. There are no short winter days with little sunlight. With constant, strong sunlight, the temperatures in the tropical rain forest remain warm throughout the year. The average temperature in the tropical rain forest is 80°F. This daily heat is required for life in the rain forest biome.

What Does *Tropical* Mean?

The word *tropical* means "having constant warm temperatures and steady rainfall that allow for year-round plant growth." Most tropical areas are found near the equator.

Obviously, a rain forest must have rain. In most of these forests, rain falls almost every day. Tropical rain forests receive an average of 160 to 400 inches of rain each year. The minimum amount of rain needed for a tropical rain forest to survive is 80 inches a year.

Tropical rain forests do not have distinct seasons. The temperatures and moisture remain constant throughout the year. There may, however, be periods of heavier and lighter rainfall.

The hot temperatures combined with lots of moisture create a very humid environment. Humidity is moisture in the air. The warm temperatures in the rain forest cause the moisture on the ground to **evaporate** quickly. As it travels up to the clouds, the moisture "hangs" in the air, causing steamy, sticky conditions. Once it reaches the clouds and cools, the moisture is returned to the ground as rain, and the cycle begins again.

The stable, year-round temperatures and moisture in the tropical rain forest allow for the gigantic plant growth and variety of animal life found in this biome.

Growing Giants

Giant plants are everywhere in the tropical rain forest. Lily pads in the Amazon Rain Forest can stretch seven feet wide. The rafflesia plant on the rain forest island of Sumatra has flowers that can weigh 38 pounds and grow 38 inches wide. They are the largest flowers in the world.

Rafflesia flower growing on the rain forest floor

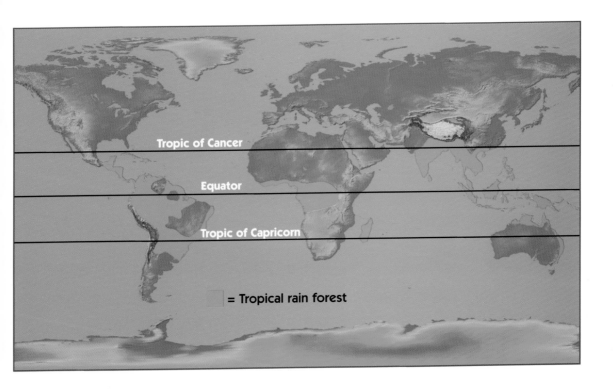

Tropic of Cancer

Equator

Tropic of Capricorn

☐ = Tropical rain forest

Where Are the Tropical Rain Forests Located?

Geographers use imaginary lines called *longitude* and *latitude* to determine where places are located in the world. The lines of latitude are used when locating tropical rain forests. The equator is a line of latitude that lies halfway between the top and bottom of the globe. It runs around the middle of the Earth like a giant belt. Tropical rain forests are found in a band that lies along the equator.

The band of rain forests stretches from the equator toward two other lines of latitude—the Tropic of Cancer to the north and the Tropic of Capricorn to the south. The part of the world that lies between these two latitude markers is called the **tropics**. The tropics have a pattern of weather that is most often hot and moist. This climate provides perfect conditions for the tropical rain forest biome.

Chapter 2

Tropical Rain Forests Around the World

Tropical rain forests cover six to seven percent of the Earth's land. The majority of these forests are found in Central and South America, Africa, and Southeast Asia.

Rain Forests of the Americas

The American rain forest stretches from eastern Mexico through Central America and into large areas of South America. Tropical rain forests can also be found in parts of Hawaii.

The largest areas of tropical rain forest are located on the continent of South America. Long ago, a huge tropical rain forest covered the land from the top to the bottom of the continent. Pieces of that large rain forest still exist, but they are now divided by pastures, roads, and towns.

The largest of all tropical rain forests is the Amazon Rain Forest in South America. Almost one-half of the world's tropical rain forest land is in the Amazon. This rain forest is approximately two-thirds the size of the United States. The huge area is home to more than one-third of all plant and animal species in the world.

North America

Mexico

Central America

South America

Rain forest

= Rain forest

Rain Forests of Africa

The second largest group of rain forests is on the African continent and the nearby island of Madagascar. This group makes up about 20 percent of the world's total tropical rain forests.

Africa is a huge continent with many different biomes. Tropical rain forests cover a small percentage of the African continent. Still, these rain forests are home to more than half of Africa's plant and animal species.

The tropical rain forests can be found in western Africa and the Democratic Republic of the Congo. The Congo River flows through much of this land. The Ituri Forest on the eastern side of the Democratic Republic of the Congo is the second largest area of tropical rain forest in the world. It stretches across 25,000 square miles of land and water.

Madagascar is the fourth largest island in the world. Millions of years ago, it split off from the African continent. Today, Madagascar lies just off the east coast of Africa.

Africa

= Rain forest

Madagascar

Visit the African Ituri in California!

The San Diego Zoo has created its own Ituri Forest. This zoo exhibit re-creates the original African rain forest **habitat**. Guests can see plants and animals of the Ituri and learn about the Mbuti people who live in the rain forests of Africa.

Fossa

The tropical rain forests of Madagascar contain many **species** of animals that are not found anywhere else in the world. Aye-ayes, fossas, and indris are all **native** to the rain forests of Madagascar. Aye-ayes are black, furry **mammals** that spend most of their time in trees. Fossas are meat-eating animals that look like a cross between a cat and a dog. Indris are large, silky, black-and-white creatures that eat leaves, flowers, and fruits.

Rain Forests of Asia

A third group of tropical rain forests is found in Southeast Asia. These forests stretch along the western coast of Burma, Thailand, and Malaysia. They also run along the eastern coast of Vietnam. The islands of Indonesia and the Philippines have tropical rain forests as well.

The rain forests of Southeast Asia have existed since the time of the dinosaurs. While the Earth has seen climate changes since that time, the Southeast Asian rain forests have been protected. These forests are surrounded by the Indian and Pacific Oceans. These bodies of water have kept the climate moist and sheltered.

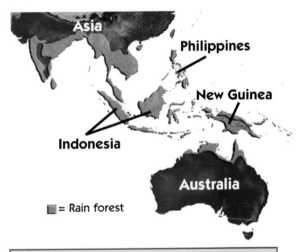

= Rain forest

Chapter 3

Layers of the Tropical Rain Forest

One of the first scientists to explore the South American rain forest was Alexander von Humbolt. In 1800, Humbolt set out to explore the Amazon. He described the Amazon jungle as "a forest above a forest." He was talking about the layers found in all tropical rain forests.

In the Jungle

A jungle is an area of thick, tangled tropical plants. It is difficult to move through these crowded patches in the rain forest because of the huge number of plants growing there.

A tropical rain forest has four layers—the forest floor, the understory, the canopy, and the emergent layer. The different layers of the rain forest form separate ecosystems. Each layer provides a different environment for plants and animals. A variety of habitats exists in each layer.

Emergent

Canopy

Understory

Forest floor

The Forest Floor

The lowest layer is the forest floor. Little sunlight reaches this layer. It is carpeted with dead leaves and fallen branches.

In the moist, hot climate of the tropical rain forest, anything that falls to the forest floor rots quickly. A leaf on the forest floor will **decompose** within two or three months. The broken-down leaf then provides **nutrients** for other plants.

Not many leafy plants grow on the floor of the rain forest. It is too dark. The **dense** layer of trees high above blocks out most of the sunlight. Only plants and **organisms** that don't need light are able to grow on the dark forest floor.

Nutrients released from decomposed plants and animals

Tapirs feast on plants on the rain forest floor. These animals can weigh up to 600 pounds.

Fungi Facts

Fungi are organisms that need little or no sunlight to grow. Fungi have no roots, leaves, or seeds. They grow on other plants or animals. Many types of fungi decompose the plants on which they grow.

Mushrooms are a type of fungi. They play an important role in cleaning the forest floor. Mushrooms break down fallen leaves into important nutrients. If they didn't, the forest floor would soon be buried in rotting waste.

are quickly absorbed by tree roots that are close to the surface. These shallow roots are able to soak up water and nutrients from the thin rain forest soil. With so much water, warmth, and nutrition, rain forest trees in higher layers grow to enormous heights.

Unfortunately, the same shallow roots that help rain forest trees grow into giants cannot always support the weight of the huge trees. Many rain forest trees overcome this problem by g r o w i n g wider. Tree trunks along the forest floor often have a triangular base that is flared out like thick wings. These large bases are called *buttress roots*. These roots help brace the gigantic trees.

Most of the large mammals in the rain forest live on the forest floor. Apes, anteaters, wild boars, antelopes, and tapirs roam the bottom layer of the forest. This layer is also home to millions of insects and spiders that feed on the nutrients on the forest floor.

Buttress roots on a kapok tree

The Understory

Above the forest floor, the tops of buttress roots are mixed in with a tangle of shrubs and small trees. This is the understory, the second layer of the tropical rain forest. Although there isn't a lot of bright light, some light does filter through to this layer.

The understory has a constant warm, humid climate. Young trees that are reaching for the bright sunlight above grow quickly in this layer. Palm trees thrive in the understory. Ferns are also common. A few ferns even grow to the size of small trees.

Some plants climb over other understory plants as they grow. The long, winding stems of lianas weave through the understory. They knit together small trees and shrubs. Lianas use tall tree trunks for support as they spiral toward the top of the understory.

Many insects live in the understory, feasting on plants and tree bark. Plants in this layer depend on these insects for **pollination**. They have flowers and fruits that attract the insects.

Small reptiles and **amphibians**, such as snakes, frogs, and lizards, make their homes in the understory. Wild cats and birds can also be found darting through this forest layer.

The Canopy

As lianas grow upward toward the sun, they break into the next layer of the tropical rain forest—the canopy. Once out of the understory, the liana grows leaves, **shoots**, and colorful flowers. In the canopy, the sun finally shines directly on the leaves and flowers of plants. The canopy plants thrive in the sunlight.

Canopy trees reach 60 to more than 100 feet tall. The trees are so closely spaced that they act as an umbrella. They shield the lower layers of the tropical rain forest from both sun and rain. When a canopy tree falls or is damaged, light is able to break through to the lower levels. This "light gap" in the canopy is quickly filled in by a **sapling** waiting in the understory.

Climbing plants find a perfect home in the canopy. Stretching up through the two bottom layers of the rain forest, these plants find the sunlight they need when they reach the canopy. The strangler fig tree is one of these plants. This tree wraps its branches around other trees to climb to the canopy. Often, the "strangled" tree dies and only the fig tree remains.

Epiphytes are another type of plant that grows in the canopy. Epiphytes are often called "air plants." Their roots

Bromeliads are short-stemmed epiphytes that often attach themselves to trunks and branches of trees. Pineapples are a member of the bromeliad family.

take in moisture from the air instead of the ground. They also get nourishment by absorbing water and minerals from the rain that falls on their leaves. Orchids and bromeliads are examples of air plants.

Sloth

Orchids All Over

More than 20,000 types of orchids grow in the tropical rain forest!

Small mammals, such as monkeys, sloths, and bats, live in the canopy. Insects, birds, and snakes weave their way through this layer. Some of them never touch the ground during their lifetimes.

The Emergent Layer

Some trees grow so large that they rise above, or emerge from, the rain forest canopy. This is the emergent layer of the rain forest and may stand 130 feet or more above the forest floor.

The top layer of the tropical rain forest receives full sunlight. The giant trees of this layer are tall and straight. Emergent trees do not grow close together. They need space for their treetops to spread. The huge branches and leaves are called the tree's *crown*. Trees in the emergent layer have gigantic crowns that spread out over the lower layers of the tropical rain forest.

Birds fly freely in the emergent layer. Monkeys jump among the branches of the gigantic trees. Butterflies and other winged insects swarm in the warm, humid air. Life is bright in the emergent layer!

Chapter 4

Life in the Tropical Rain Forest

Imagine a day in the tropical rain forest. Large green leaves rustle in the breeze. Dead leaves and branches crunch underfoot. Birds chirp overhead. Insects buzz loudly. People work together in communities to make use of the forest's resources. The steamy rain forest is bursting with life!

Only six to seven percent of the Earth is covered with tropical rain forests. But this small area of land contains more than half of all plant and animal species in the world!

Rain Forest Plants

Forests are made up of different species of trees. A two-and-a-half-**acre** area of tropical rain forest can have up to 750 types of trees growing on it. Palm, teak, cypress, and mangrove trees are common in the rain forest. Huge kapok trees are home to many animals. Hardwood trees, such as mahogany, grow tall and strong. Rubber trees produce latex, which is used to make rubber. Pods from the cacao tree are used to make chocolate. Fruit trees provide food for animals and people of the forest. Oranges, lemons, berries, coconuts, bananas, star fruits, guavas, and mangoes are plentiful.

Over 1500 kinds of plants may be found in the same two and a half acres of rain forest. Colorful flowers, such as hibiscus, periwinkle, and violets, brighten the forest.

Bananas are one of the top food products sold around the world. The majority of these bananas are grown in tropical rain forests.

Many rain forest plants are a source of foods and spices around the world. Everyday foods, such as coffee, sugar, vanilla, pineapples, and avocados, are products of the rain forest. Peanuts, cashews, and Brazil nuts are rain forest treats. Cinnamon, nutmeg, ginger, and pepper are just a few spices that come from tropical rain forest plants.

Many of the medicines used today have been developed from plants found in the rain forest. Often, by watching how native rain forest people use plants, scientists are able to make drug discoveries. At least one-fourth of all medicines used today come from rain forest plants.

Rain Forest Animals

Over half of the wildlife in the world calls the tropical rain forest its home. Spider monkeys swing among the lianas. Poisonous viper snakes slither through the trees, hunting for food. Toucans with orange and green beaks fly through the air.

Each layer of the tropical rain forest provides a different ecosystem for animals. Some creatures spend their entire lives in just one layer of the rain forest. Others are able to survive in more than one layer.

Jungle cats of the tropical rain forest are top **predators**. They hunt for **prey** in the rain forest. While these wild cats are excellent climbers, they often hunt larger, nonclimbing animals on the forest floor. Jaguars and ocelots live in the understory of Central and South American rain forests. Their spotted coats help camouflage them as they stalk wild pigs. The leopards of the African forests might make a meal of the okapi, a large **herbivore** with zebralike markings.

The okapi is also known as the "forest giraffe." It looks like a small giraffe with a short neck and has a sticky tongue to collect leaves from understory trees.

Apes are another group of mammals that move between the rain forest layers. Gorillas of West Africa are large, heavy apes that spend most of their time on the forest floor. At night, they climb to the lower understory to sleep in a nest of tree branches.

Orangutans are also large apes. An adult orangutan may be 5 feet tall and weigh over 200 pounds. But unlike gorillas, orangutans of Borneo and Sumatra spend time in the rain forest canopy. By swinging hand over hand, orangutans move gracefully through the high branches. At night, the orangutans build a nest of branches and leaves up in the tall trees.

Jaguar

Above the sleeping gorillas, African gray parrots and giant swallowtail butterflies glide through the rain forest canopy. Brown fruit bats collect fruit from canopy trees, scattering seeds as they eat. Some of the seeds fall to the forest floor where they sprout into new plants.

Some unusual animals fly high in the tropical rain forest canopy. Flying lizards and flying lemurs are found in the rain forests of Asia. The "wings" of these flying animals are actually large flaps of skin. When gliding from tree to tree, the animals spread their flaps like wings.

Only the best climbers and flyers make it to the emergent layer of the tropical rain forest. Colobus monkeys perch in the tallest trees. Large birds of prey, such as eagles, soar among the emergent trees.

Insect Invasion

More than 50 million different insects creep, crawl, fly, and inch their way through the tropical rain forest!

Rain Forest People

While animals and plants fill the upper layers of the tropical rain forest, people of the rain forest live on the forest floor. Native people have lived in tropical rain forests for thousands of years. Long ago, the number of people living there was large. Today, the number has decreased.

As many as five million people used to live in the Amazon Rain Forest in South America. In the last 100 years, over 90 different Amazonian tribes have become **extinct**. Today there are fewer than 250,000 of Brazil's native Amazon people left.

Rain forest people live among a wealth of plants and animals. This means that many rain forest tribes do not grow their own food. They just move from place to place in search of the food nature has provided. They hunt animals and gather wild fruits and vegetables. These groups are called *hunter-gatherers.*

The Mbuti people often paint their faces as a form of art and as decoration for special ceremonies.

In the Ituri Forest of Africa, the Mbuti tribe lives as hunter-gatherers. The tribe roams the rain forest, living in groups of about 30 people. Sometimes the tribe camps in one location. If conditions are right, the Mbuti may stay for a few weeks or months. During this time, the Mbuti women build huts using materials gathered from the rain forest. Large leaves cover the huts and provide shelter from the rain.

Not all rain forest natives are hunter-gatherers. Some tribes grow small amounts of food using a method known as slash-and-burn agriculture. A tribe clears a small area of rain forest by slashing down all the trees and shrubs. Anything left in the area is then burned. After the fire dies down, the ashes serve as **fertilizer** for new crops. After a few years, the nutrients in the soil are gone. The tribe then leaves the area and starts over. Once the tribe moves on, the tropical rain forest regrows over the farmed spot.

About ten thousand Yanomami live in northern Brazil and southern Venezuela. This group lives in villages scattered throughout the rain forest. They grow some of their food in small fields created by slash-and-burn agriculture.

Chapter 5

Protecting the Treasures of the Tropical Rain Forest

The tropical rain forest has many treasures—animals, plants, products, and people. Like all biomes, the rain forest has a delicate balance of habitats and ecosystems. With so many people using the land and products, some of that balance has been destroyed. As a result, the tropical rain forest has become a threatened environment.

Deforestation

Deforestation is the removal of many trees in an area. This leaves the forest floor bare to heavy tropical rains. The pounding rains quickly wash away the thin soil. Once the soil washes away, plants find it nearly impossible to regrow on the forest floor.

Soil washed away from the forest floor flows into the rivers. Over time, the soil builds up in the riverbeds. The rivers become blocked. Blocked riverbeds cause flooding, which further disrupts the rain forest ecosystems.

Deforestation damages plant habitats of the tropical rain forest. As a result, animals and native people lose their habitats as well.

Deadly Deforestation

Deforestation of the world's tropical rain forest occurs every day. In fact, about 6000 acres of rain forest disappear every hour!

Loggers clear an area of the rain forest in West Kalimantan, Borneo.

Reasons for Deforestation

Deforestation occurs for many reasons. Slash-and-burn agriculture is the clearing of rain forest land for cattle and crops. When small groups of native people work with hand tools to clear small patches of land, the rain forest is usually able to recover. But when slash-and-burn agriculture is practiced by a large number of people working with large machinery, the tropical rain forest is permanently destroyed. Huge areas of rain forest are cleared for growing crops or grazing cattle. This land is never **recovered**.

Deforestation also occurs when trees in the tropical rain forest are cut down for lumber. Hardwoods, such as mahogany and teak, are in great demand for furniture and boats. Countries with rain forest land where these trees grow sell them for large profits. Other softer trees are also cut down and sold to make paper and **plywood**.

Many countries depend on money made from the products that are harvested from the tropical rain forests. Medicines, foods, rubber, bamboo, and other resources from the rain forests are a large source of income for these countries. But removing these products from the forests can cause damage beyond repair.

Deforestation Damages
Climate

The gigantic trees of the tropical rain forest soak up large amounts of rainwater through their roots. This water travels upward through the trees. It is later released through the leaves as water vapor, or steam. The released water vapor forms clouds. The same water then falls from the clouds as rain. This process of water moving from the clouds to the trees and back again is known as the water cycle.

When the trees of the tropical rain forests are cut down, the water cycle is broken. Without the rain forest plants to store water, the air above the rain forest becomes drier. Less rain may mean **droughts**. Temperatures may increase due to lack of water and tree cover. If too much rain forest land is destroyed, rainfall and temperature patterns all over the world may be affected.

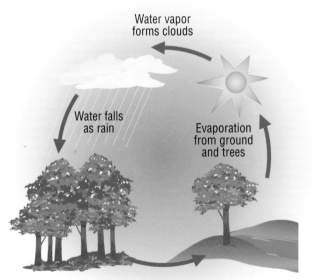

Water vapor forms clouds

Water falls as rain

Evaporation from ground and trees

The Water Cycle

Heating Up

Plants take in and store carbon dioxide. When they are burned to clear rain forest land, large amounts of carbon dioxide are released into the air. Carbon dioxide traps heat in the air. This is called the *greenhouse effect*. Over time, this can cause a rise in temperatures on the planet.

Plants and Animals

The tropical rain forest is home to more plant and animal species than any other biome. When parts of the rain forest are destroyed, habitats disappear. This causes the loss of many rain forest creatures.

Large animals, such as leopards, use large amounts of the tropical rain forest as their habitat. They cannot exist in small forests. Insects that rely on food from the leaves or bark of trees disappear when the trees are cut down. Hummingbirds and songbirds from colder lands **migrate** to the warm rain forests for the winter. As the tropical rain forests are deforested, these huge flocks of birds arrive each winter to find their habitats destroyed. Without shelter and food, the bird population decreases.

Every day, an average of 137 different types of plants and animals are lost forever from the tropical rain forest.

Native People

The native people of the rain forest also suffer from deforestation. Their lands and lifestyles are threatened by

Leopards race across the forest floor but are also good tree climbers. These clever cats often hide their food in trees.

modern people, such as miners, ranchers, and logging companies.

The Penan are a hunter-gatherer tribe that lives in the Asian rain forest of Borneo. When logging companies came to Borneo, they found a land almost entirely covered in tropical rain forest. The companies quickly started cutting down the trees of the rain forest where the Penan hunted.

Protests and complaints by the Penan did no good. Their habitat has almost been completely destroyed. About 7000 Penan remain today, but only about 300 are still able to live as hunter-gatherers.

Hope for the Tropical Rain Forest

Some people are trying to slow down the fast rate of deforestation. Stopping the destruction of the world's tropical rain forests will take all people working together. But with so many people depending on the rain forests, it's not easy to find solutions that suit everyone.

Countries that earn money from tropical rain forest lumber need help finding new sources of income. Programs have been started to grow **seedling trees** to replace some of the trees harvested from the rain forest. Tree nurseries in central Africa raise fast-growing species of trees for local people to use instead of cutting down rain forest trees.

Governments can help by setting aside part of the rain forest as national parks. Manu National Park in Peru was declared a national park in 1973. Besides saving the plants and animals, Manu National Park provides a safe home for native tribes.

While most people eat products from the rain forest, it is important that these resources are removed without damaging the biome. Many foods are now labeled "sustainably harvested from the rain forest." That means that the food came from the rain forest without harming any plants, animals, or native people.

Tourism has become a new way to save the rain forests. Visiting tourists bring much-needed money to rain forest countries. Visitors are able to see the beauty of the rain forest and its wildlife for themselves.

Chew on This!

Thousand of acres of rain forest land are cleared each year for beef cows to graze on. Did you know that 55 square feet of rain forest is destroyed for every quarter-pound hamburger that comes from the rain forest? That's a lot of lost rain forest for a burger!

Orangutans spend most of their time swinging from the branches of rain forest trees. They eat the leaves, fruit, seeds, and flowers from the trees and drink water that collects in between branches.

In Borneo, tourists can visit an orangutan breeding center to see these endangered animals up close.

At one time, 6 million square miles of tropical rain forest existed in the world. Today, only 2.6 million square miles remain. The world must work together to find a balance between using resources from the rain forest and protecting this valuable biome.

Internet Connections and Related Reading for Tropical Rain Forests

http://www.animalsoftherainforest.
org/map.htm
Check out a map of the tropical rain
forests, and find a list of the countries
where they are located. Read about
reasons to save the rain forests, and
explore the list of rain forest links!

http://www.ran.org/kids_action/
Visit the Kids Corner of the Rainforest
Action Network. This site has lots of
information on tropical rain forests,
actions to take to protect this biome,
crafts and recipes, and even a rain
forest slide show.

http://www.EnchantedLearning.com/
subjects/rainforest/
Learn all about the rain forest at this
site. Discover the rain forest's layers,
locations, and animals. Try some of the
fun activities.

http://mbgnet.mobot.org/sets/rforest/
plants/valley.htm
Examine the plants of the tropical rain
forest—their adaptations and
characteristics and the products the
world gets from them.

http://worldforest.geo.msu.edu/rfrc/
tour/rainforest.html
Take a tour of a tropical rain forest in
Nicaragua. See some of the unique
plants, animals, and people of this biome.

....

At Home in the Rain Forest by
Diane Willow. A direct, descriptive
text accompanies this lushly visual
journey through the rain forest, making
observations about the flora and fauna,
from the emergent layer through the
upper and lower canopies.
Charlesbridge Press, 1991.
[RL 3 IL 1–4] (4456301 PB
4456302 CC)

Canopies in the Clouds: Earth's Rain Forests by Ellen Hopkins. Take a journey to the Amazon Rain Forest, and learn about the plants and animals that live there. Find out what is being done to save this natural resource. Perfection Learning Corporation, 2002. [RL 4.8 IL 4–9] (3243101 PB 3243102 CC)

Nature's Green Umbrella: Tropical Rain Forests by Gail Gibbons. With clear text and colorful illustrations, young readers see what makes rain forests so special and why people are working so hard to save them. William Morrow, 1997. [RL 2 IL K–4] (5519801 PB)

One Day in the Tropical Rainforest by Jean Craighead George. Today is doomsday for a young Venezuelan Indian boy's beloved rain forest and its animal life—unless he and a visiting naturalist can save it. HarperCollins, 1995. [RL 4 IL 3–7] (4801001 PB 4801002 CC)

Rain Forests: A Nonfiction Companion to Afternoon on the Amazon by Will Osborne and Mary Pope Osborne. This research guide is Jack's and Annie's very own guide to the rain forests. Includes up-to-date information, exciting photos, fun illustrations, interesting tidbits, and much more. Random House, 2000. [RL 2.8 IL K–4] (3272701 PB)

What Is a Biome? by Bobbie Kalman. This book introduces biomes, showing and describing the main kinds and discussing their location, climate, and plant and animal life. Crabtree Publishing, 1998. [RL 3 IL 2–5] (5729401 PB 5729402 CC)

• RL = Reading Level
• IL = Interest Level
Perfection Learning's catalog numbers are included for your ordering convenience.
PB indicates paperback. CC indicates Cover Craft. HB indicates hardback.

Glossary

acre (AY ker) area equal to 43,560 square feet

adapt (uh DAPT) to learn to successfully live in an environment (see separate entry for *environment*)

amphibian (am FIB ee uhn) scaleless animal that lives in moist environments on land and in water (see separate entry for *environment*)

biome (BEYE ohm) environment with unique features (see separate entry for *environment*)

camouflage (CAM uh flahzh) to hide in one's environment (see separate entry for *environment*)

climate (KLEYE mit) pattern of weather

community (kuh MYOU nuh tee) organisms that live together in a particular location (see separate entry for *organism*)

coral reef (KOR uhl reef) rocky area in warm, shallow ocean waters created from the remains of animals called *polyps*

decompose (dee com POHZ) to break down into smaller parts

dense (dens) crowded; very thick

drought (drowt) period of time with little or no water

ecosystem (EE koh sis tuhm) group of living creatures that interact with one another and their surroundings

environment (en VEYE er muhnt) set of conditions found in a certain area; surroundings

equator (ee KWAY ter) line of latitude that lies halfway between the top and bottom of the globe

evaporate (ee VAP or ayt) to change from a liquid to a gas

extinct (EK stinkt) died out; gone forever

fertilizer (FER tuh leye zer) material that increases a soil's ability to support plant life

habitat (HAB i tat) place where a plant or animal lives

herbivore (ER buh vor) animal that eats plants

mammal (MAM uhl) animal that is covered with hair and feeds its young with milk from the mother

marsupial (mar SOO pee uhl) mammal whose females have a pouch on the stomach to carry their young (see separate entry for *mammal*)

migrate (MEYE grayt) to move from one area to another

native (NAY tiv) originally living or growing in an area

nutrient (NOO tree ent) material that living things need to live and grow

organism (OR guh niz uhm) living thing

plywood (PLEYE wood) sheets of wood glued together that are used as a building material

pollination (pah luh NAY shuhn) transfer of pollen, a material needed to create new plants

predator (PRED uh ter) animal that hunts other animals for food

prey (pray) animal that is hunted by other animals for food

recovered (ree KUH verd) brought back to normal conditions

reptile (REP teyel) animal with scales that crawls or moves on its belly or short legs

sapling (SAP ling) small tree

seedling tree (SEED ling tree) tree grown from planted seeds

shoot (shoot) stem or branch that isn't fully grown

species (SPEE sheez) group of living things with common characteristics that can reproduce with one another

trench (trench) deep canyon, or valley, on the ocean floor

tropics (TROP iks) area of the world that stretches from the equator toward the Tropic of Cancer to the north and the Tropic of Capricorn to the south

tundra (TUHN druh) treeless region with soil that is often frozen year-round

underbrush (UHN der bruhsh) shrubs, bushes, and small trees growing under larger trees

Index